NADIA SAFIA CHENOUF
CHAFIK REDHA MESSAI
ISABEL CARVALHO

Mechanisms of bacterial resistance to antibiotics

NADIA SAFIA CHENOUF
CHAFIK REDHA MESSAI
ISABEL CARVALHO

Mechanisms of bacterial resistance to antibiotics

ScienciaScripts

Imprint

Any brand names and product names mentioned in this book are subject to trademark, brand or patent protection and are trademarks or registered trademarks of their respective holders. The use of brand names, product names, common names, trade names, product descriptions etc. even without a particular marking in this work is in no way to be construed to mean that such names may be regarded as unrestricted in respect of trademark and brand protection legislation and could thus be used by anyone.

Cover image: www.ingimage.com

This book is a translation from the original published under ISBN 978-620-3-45885-5.

Publisher:
Sciencia Scripts
is a trademark of
Dodo Books Indian Ocean Ltd. and OmniScriptum S.R.L publishing group

120 High Road, East Finchley, London, N2 9ED, United Kingdom
Str. Armeneasca 28/1, office 1, Chisinau MD-2012, Republic of Moldova, Europe
Printed at: see last page
ISBN: 978-620-6-07905-7

MECHANISMS OF BACTERIAL RESISTANCE TO ANTIBIOTICS

NADIA SAFIA CHENOUF[1,2,3] CHAFIK REDHA MESSAI[2,4] ISABEL CARVALHO[5]

1Laboratory of Exploration and Valorization of Steppe Ecosystems (EVES), Department of Biology, Faculty of Nature and Life Sciences, University of Djelfa, Djelfa, Algeria.

2Faculty of Nature and Life Sciences and of Earth and Universe Sciences, University Mohamed El-Bachir El-Ibrahimi of Bordj Bou Arréridj, Algeria.

3Laboratoire de Biologie des Systèmes Microbiens (LBSM), BP92, 16050, Ecole Normale Supérieure de Kouba, Algiers, Algeria.

4Laboratoire de Santé et Productions Animales, Rue Issad Abbes, Oued Smar 16000, Ecole Nationale Supérieure Vétérinaire, Algiers, Algeria.

5Microbiology and Antibiotic Resistance Team (MicroART), Department of Veterinary Sciences, University of Trás-os-Montes and Alto Douro, 5000-801 Vila Real, Portugal.

TABLE OF CONTENTS

1. HISTORY AND LANDMARKS CHRONOLOGY

The era of antibiotics began in 1928 with the accidental discovery by Scottish scientist Alexander Fleming (1881-1955) of the ability of the mold Penicillium notatum to inhibit the growth of the bacterium Staphylococcus aureus. At first, however, Fleming's work was confined to the relatively small circle of bacteriologists and biochemists, who at the time were much more inclined than physicians to accept microbial antagonism as a potential source of therapeutic agents. It was only ten years later that Fleming's work was taken up by the Australian pathologist Howard Walter Florey (1898-1968) and the German biochemist Ernst Boris Chain (1906-1979), who were able to master the instability of penicillin and succeeded in purifying it, demonstrating its therapeutic efficacy in staphylococcal infections in 1941. From then on, the pharmaceutical industry began a mad scramble to find new compounds with the same antibacterial activity. Between 1941 and 1950, the first representatives of the main antibiotic families were discovered: streptomycin, chloramphenicol and tetracycline. These new molecules broadened the spectrum of antibiotic activity, improving the fight against bacterial diseases (Desiderio, 1954; Helfand, 1982; Guillot, 1989).However, the golden age of antibiotics, as predicted by evolutionary theory, did not last long. Bacteria, like any other living organism, evolve when selective pressure is introduced (e.g. antibiotics), given the conditions of time, heredity and variability. The widespread and intensive use of antibiotics both in human medicine and in agriculture and animal husbandry has accelerated the appearance, selection and spread of bacteria resistant to these compounds (Alonso, 2018). Penicillinase-producing bacteria were first described shortly after the discovery of penicillin. By the late '50s, 80% of Staphylococcus aureus isolated from hospitalized patients were resistant to penicillin, due to the presence of penicillinase-producing bacteria. the production of β-lactamases. This led, in the early 1960s, to the development by semi-synthesis of penicillinase-stable penicillin derivatives (methicillin, cloxacillin and oxacillin). Similarly, resistance to methicillin was observed less than a year after its clinical introduction. In 1955, during an epidemic of bacillary dysentery, isolated Shigella strains, initially susceptible, simultaneously became resistant to streptomycin, chloramphenicol, tetracycline and sulfonamides. In 1960, it was demonstrated that this multiple resistance had been transferred to Shigella in the intestines of patients by simple contact with multi-resistant Escherichia coli. These observations led geneticists to speculate that resistance was associated with an extra-chromosomal structure they called a "plasmid" (Guillot, 1989; Fong et al., 2019). During the 1970-1980 decade, new stable β-lactams (notably extended-spectrum cephalosporins) were developed. However, their excessive clinical use was followed by the early emergence of resistance. Thus, the first β-lactamase capable of hydrolyzing these new molecules was described in 1985 in a Klebsiella pneumoniae strain from Germany (Bradfort, 2001; Paterson and Bonomo, 2005). In 1986, resistance to vancomycin was detected in Enterococcus isolates in France and the UK (Leclercq et al., 1988; Uttley et al., 1989). The figure below summarizes the history of antibiotic discovery and the parallel evolution of antibiotic resistance.

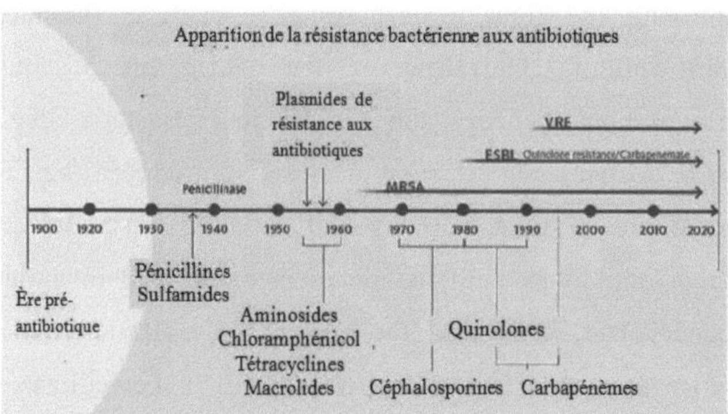

Figure 1: Discovery of antibiotics and evolution of resistance (Fong et al., 2019)

MRSA: Methicillin-resistant S. aureus; **VRE**: Vancomycin-resistant Enterococci; **ESBL**: extended-spectrum β-lactamases.

2. GENETIC SUPPORTS FOR THE ACQUISITION AND DISSEMINATION OF RESISTANCE GENES

Bacterial resistance to antibiotics is a relative concept. Indeed, a large number of definitions have been suggested, depending on the field of study: clinical, epidemiological, biochemical, microbiological or genetic. However, the most common definition is based on microbiological (in vitro resistance) and clinical (in vivo resistance) criteria (Muylaert and Mainil, 2013).

This resistance can have two origins:

- **Intrinsic (natural) resistance: This is** a functional or structural characteristic forming part of the genetic heritage and conferring a certain tolerance, or even total insensitivity, to members of the same bacterial species or genus (Courvalin, 2008; Muylaert and Mainil, 2012). The use of antibiotics in this case implies the selection of these resistant populations, thus favoring their spread;

- **Extrinsic (acquired) resistance:** This phenomenon appears in a member(s) of an initially susceptible bacterial species, through the acquisition of mechanisms not existing in the normal bacterial population. Two major processes underpinning this genetic evolution are described: chromosomal mutations and horizontal gene exchange, which can occur in a variety of ways, including transformation, transduction and conjugation (figure 2). In the latter process, a series of genetic carriers for the acquisition and dissemination of resistance genes may come into play, namely plasmids, integrons and transposable elements (Furuya and Lowy, 2006; Alekshun and Levy al., 2007).

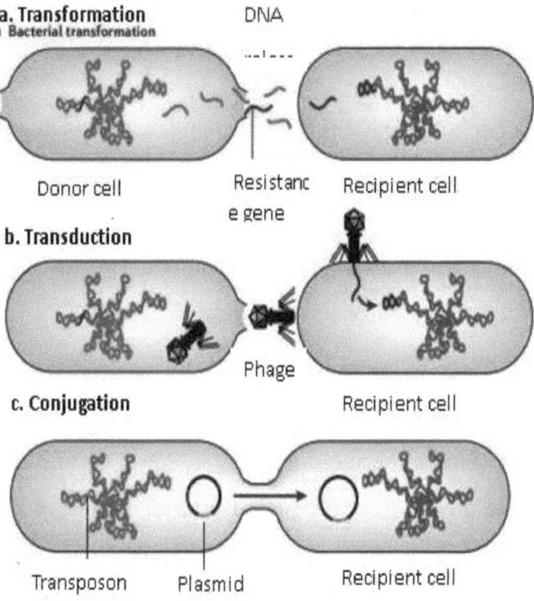

Figure 2: Mechanisms of horizontal transfer of resistance genes between bacteria(Furuya and Lowy, 2006)

2.1. Plasmids

A plasmid is an extra-chromosomal, double-stranded, circular DNA molecule with an origin of replication that enables it to replicate autonomously. It does not contain genes essential to the survival of the host cell, but rather genes that confer an adaptive advantage, notably genes controlling antibiotic resistance and virulence factors. Plasmids acquire these genes through mobile genetic elements: insertion sequences (IS) and transposons (Couturier et al., 1988; Carattoli, 2013). The relationship between the various mobile genetic elements is illustrated in Figure 3.

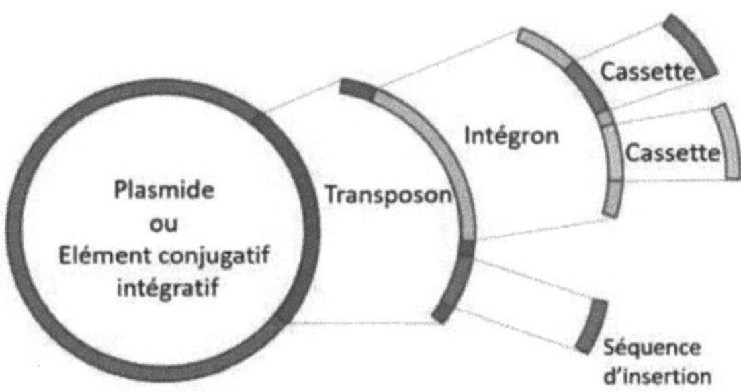

Figure 3: Simplified diagram of the relationship between the various mobile genetic elements (Couvé-Deacon, 2017).

Many resistance plasmids are conjugative, i.e. they encode the functions needed to promote their own transfer from one bacterial cell to another (self-transmissible). Others are non-conjugative, but can be mobilized when assisted by a conjugative plasmid co-existing in the cell. As a result, non-conjugative plasmids tend to be relatively small, often less than 10 kb, while conjugative plasmids can be over 30 kb in size (Bennet, 2008). In 1971, a plasmid classification scheme based on plasmid stability during conjugation was proposed. This phenomenon is called "plasmid incompatibility" (Datta and Hedges, 1971). This refers to the inability of two plasmids to co-exist stably in the same cell line. The first incompatibility groups were defined as follows: IncI, plasmids producing type I pili sensitive to phage Ifl; IncN, plasmids related to N3 sensitive to phage IKe; IncF, plasmids producing type F pili sensitive to phage Ff; and IncP, plasmids related to RP4 sensitive to phage PRR1 (Carattoli, 2008).The smallest part of a plasmid that replicates with the same characteristics as the parent plasmid is called a "replicon" (Kollek et al., 1978). Each replicon contains: an ori origin of replication, genes encoding proteins required for the origin of replication (Rep) and proteins that bind to ori and regulatory factors (Couturier et al., 1988, Kollek et al., 1978). On this basis, Couturier et al (1988) developed a method for typing Enterobacteriaceae plasmids, based on hybridization of replicons of the different incompatibility groups. Eight years later, the IncP, IncN, IncW and IncQ groups were identified (Götz et al., 1996).

In 2005, Carattoli et al. (2013) designed 18 primer pairs with five multiplex PCRs and 3 single PCRs; as a result, the major incompatibility groups in Enterobacteriaceae were recognized: FIA, FIB, FIC, HI1, HI2, I1, L/M, N, P, W, T, A/C, K, B/O, X, Y, FII and FIIA.This plasmid typing method is known as inc/rep PCR-based replicon typing PBRT. At least 27 incompatibility groups have been identified in enterobacteria. However, there are a number of sequenced plasmids that could not be assigned to any known Inc group (Shintani et al., 2015; Wang et al., 2019).

2.2.Integrate

Discovered relatively recently (late 1980s), integrons are considered to be the most efficient genetic element known to the bacterial world for the dissemination of antibiotic resistance genes (Skurnik, 2009). They have been widely studied in Gram-negative bacteria (Jouini et al., 2007; Jones-Dias et al., 2016), but have also been demonstrated in Gram-positive bacteria such as Corynebacterium glutamicum (Nesvera et al., 1998) and Staphylococcus aureus (Xu et al., 2011). Integrons are a system for capturing and expressing exogenous coding regions, generally without promoters, known as "cassettes". The three fundamental functional elements of the integron are (Guérin et al., 2011):

- the gene (intI) encoding an integrase responsible for acquiring or excision of cassettes by site-specific recombination;

- the recombination site (attI) where cassette-gene integration occurs;

- the Pc promoter for expression of integron cassette genes. Sometimes integrons have a second, stronger promoter, located adjacent to the 3' position of the first, which increases the degree of transcription and gene expression (Figure 4).

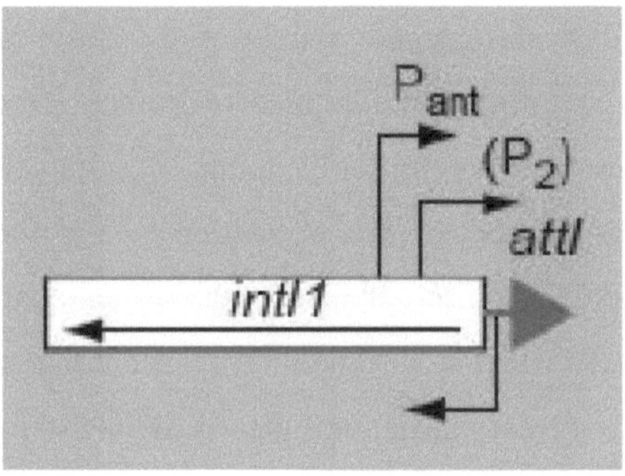

Figure 4: Basic structure of an integron (Sabaté and Prats, 2002)

Currently, there are two types of integrons: resistance integrons (RIs), generally carried by plasmids or transposons, and chromosomal integrons, also called super-integrons (SIs) because they contain a large number of cassettes (between 20 and 200 cassettes with mostly unknown functions) (Guérin et al., 2011). The amino acid sequence of the integrase makes it possible to differentiate nine classes of integrons. So far, five classes (intI1, intI2, intI3, intI4 and intI5) are linked to insertion sequences, transposons and conjugative plasmids and may be implicated in the spread of resistance between species, or within the same species. For this reason, they are referred to as "mobile resistance integrons" (Mazel et al., 2006). Class 1 integrons are associated with functional and non-functional transposons derived from Tn402, which can be incorporated into long transposons such as Tn21 (Brown et al., 1996). Class 2 integrons are derived from the Tn7 transposon (Radström et al., 1994) and class 3 integrons may be localized in a transposon located in an as yet uncharacterized plasmid (Collis et al., 2002). The other two classes of mobile integrons, 4 and 5, have been identified in relation to trimethoprim resistance in Vibrio, since the class 4 integron is part of an SXT element found in Vibrio cholerae (Hochhut et al., 2001), and the class 5 integron is localized in a compound transposon in a Vibrio salmonicida plasmid (Mazel et al., 2006). Class 6, 7, 8 and 9 integrons were detected only once each, from strains isolated in the environment: class 6 and 7 with a cassette gene of unknown function (Clark et al., 2000; Nield et al.,

2001), class 8 without cassette genes (Nield et al., 2001) and class 9 with the dfrA1 cassette gene encoding trimethoprim resistance (Hochhut et al., 2001). Class 1 integrons are most commonly found in clinical strains. They are characterized by conservation of the 5' sequence, and most in turn contain another conserved 3'CS region which contains a quaternary ammonium resistance gene (qacΔE1) and a sulfonamide resistance gene (sul1) (figure 5). These two genes are not cassettes; they are fixed in the integron (Sabaté and Prats, 2002).

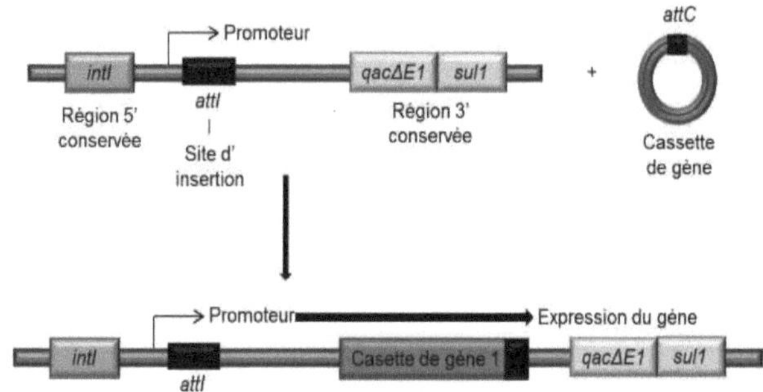

Figure 5. Structure of a class 1 integron and representation of the capture and expression mechanism of a gene cassette (Muylaert and Mainil, 2013).

2.3. Transposing

Discovered by Barbara McClintock in 1951 during her work on genetic instability in maize, transposable elements are DNA sequences that have the property of changing position in the genome (the chromosome or a plasmid); they are also known as "mobile genetic elements". The movement performed by the transposon is called "transposition", and the enzymes responsible for it are called "transposases". The transposon, also nicknamed the "jumping gene", differs from the integron in that it encodes its own transposases and retains its ability to "jump" from one region of the genome to another as it moves. Thus, all bacterial transposons are flanked at their ends by inverted, repeated DNA sequences, the targets of the transposases, and by short, repeated sequences of the target DNA that flank the transposon (Vinué, 2010; Muylaert and Mainil, 2012). There are three types of transposon (Figure 6): insertion sequences (IS), composite transposons and non-composite (complex) transposons.

2.3.1. Insertion sequences (IS)

These are the simplest transposable elements. Generally less than 2500 bp in size, they are phenotypically cryptic, encoding only the genetic information required for transposition. They consist of two short direct target DNA sequences (DR) of 2 to 14 bp and two inverted repeat DNA sequences (IR) of 20 to 40 bp, the targets of transposases, flanking a central portion formed by the gene encoding the transposase (Merlin and Toussaint, 1999; Muylaert and Mainil, 2013).IS are involved in a large number of deletions, inversions and reorganizations of the bacterial genome, affecting its assembly and the clustering of genes with adaptive functions such as antibiotic resistance (e.g. IS26), virulence and even catabolic functions (Haren et al., 1999; Bennet, 2004; Varani et al., 2021).Since their discovery in the late 60s, the number and diversity of IS has continued to grow. At present, the sequences of over 4,000 different IS are deposited in the ISfinder database (https://www- is.biotoul.fr/). IS are classified into at least 27 families (Siguier et al., 2015; Varani et al., 2021).

2.3.2. Transposons composites

Composite transposons consist of two insertion sequences, generally oriented in opposite directions, flanking a central portion comprising antibiotic resistance gene(s) (Muylaert and Mainil, 2012). Sometimes the IS elements are identical, as in the case of transposon Tn9 (direct repeats of IS1, encoding resistance to chloramphenicol) and Tn903 (inverted repeats of IS903, encoding resistance to kanamycin) (Alton and Vapnek, 1979; Grindley and Joyce, 1980). On other occasions, as in Tn5 or Tn10, the IS elements are linked, but not identical (Haniford, 2006).

2.3.3. Non-composite transposons (complex)

With a more complex structure than the previous two types, and of greater clinical importance due to its extreme efficiency in the dissemination of antibiotic resistance genes, non-composite transposons are characterized by the absence of IS at their ends. Most of t h e m   are related by their transposase and by their 35 to 48 bp terminal sequences, repeated in reverse orientation and recognized by the transposase. These transposons vary in size (up to 70 kb in the case of Tn4651). They contain genetic information essential for transposition and so-called "auxiliary" information, which may be catabolic genes (Tn4651) or antibiotic resistance genes (Tn3, Tn21 and Tn1721) (Merlain and Toussaint, 1999; Bennet, 2008).

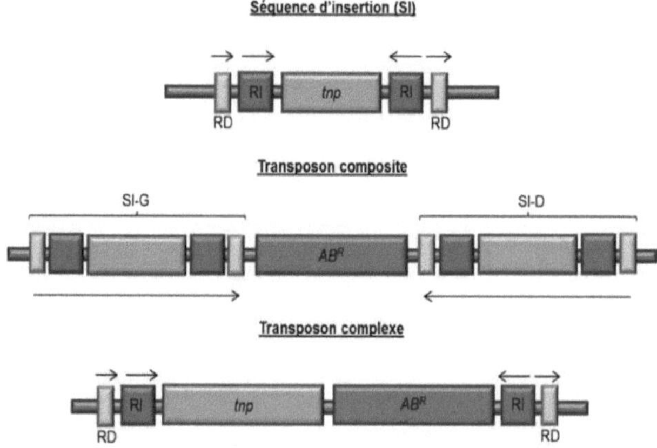

Figure 6. Simplified representation of the three types of transposons (Muylaert and Mainil, 2013).

3. MECHANISMS OF BACTERIAL RESISTANCE TO ANTIBIOTICS

For an antibiotic to be able to inhibit bacterial growth, it must first cross the cell wall and reach its target at sufficient concentrations without being rejected or metabolized. Antibiotics act at different levels, blocking certain metabolic processes essential to bacterial survival: cell wall synthesis, cytoplasmic membrane synthesis, protein or nucleic acid synthesis (replication, transcription and translation) and folate synthesis. However, certain bacteria can neutralize the action of these antibacterials by several mechanisms involving modification of the target (mutation, variability), of the antibiotic molecule itself (enzymatic inactivation) and modification of the intracellular concentration of the antibiotic (impermeability and efflux systems) (figure 7) (Pagès, 2004):

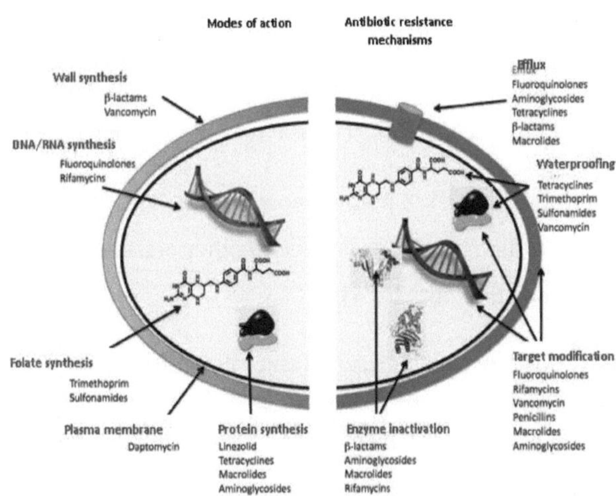

Figure 7. Main modes of action of antibiotics and resistance mechanisms (Wright, 2010)

3.1. Resistances to β- lactamins

In antibiotic therapy, the β-lactams (penicillins, cephalosporins, monobactams, carbapenems and β-lactamase inhibitors) are the most important antibiotic family, both in terms of the number and diversity of molecules available and their therapeutic indications. The β-lactam antibiotics are characterized by the constant presence of the β-lactam ring, combined with variable rings and side chains. These bactericidal molecules exert their activity by inhibiting the final stage of cell wall synthesis. These antibiotics act as structural analogues of the natural substrate (alanine dipeptide) for transpeptidases and carboxypeptidases, and bind to specific PLPs (penicillin-binding proteins) in the bacterial cell wall to inhibit the formation of bridges between peptidoglycan chains (Bryskier, 1999; Suárez and Gudiol, 2009). In Gram-negative bacilli, three mechanisms of β-lactam resistance have been described (Figure 8):

1- impermeability due to loss or alteration of porins (mainly OmpF and OmpC) or outer membrane proteins (Gutmann, 1986; Pagès et al., 2004);

2- expulsion of antibiotics by efflux pumps, notably the AcrAB-TolC and AcrAD-TolC systems, which confer resistance to other classes of antibiotics (Sun et al., 2014; Kobayashi et al., 2014);

3- enzymatic inactivation by β-lactamases (Tang et al., 2014).

Resistance to β-lactam antibiotics in staphylococci is based on two main types of mechanism (figure 8):

1- an extrinsic resistance mechanism: production of degrading enzymes antibiotic(Tang et al., 2014);

2- an intrinsic resistance mechanism: modification of PLPs or acquisition of newPLPs (Gutmann, 1986).

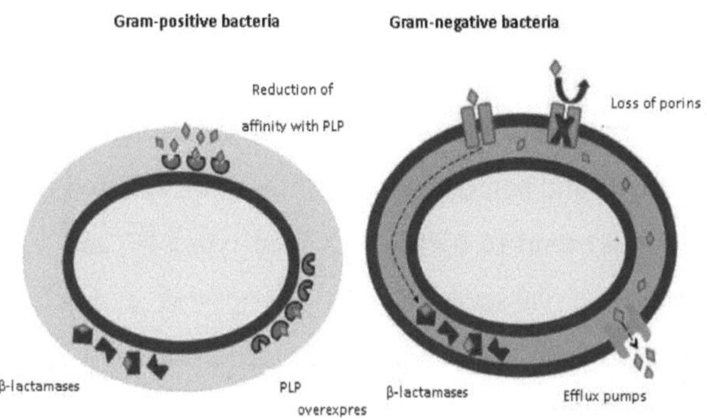

Figure 8. Main mechanisms of β-lactam resistance in Gram positive andGram-negative (Tang et al., 2014)

In staphylococci, enzymatic inactivation is due to the action of penicillinases encoded by the blaZ gene, which may be carried either by a transposon or be chromosomal (Quincampoix and Mainardi, 2001). In 2018, whole-genome sequencing of a penicillin-resistant Staphylococcus arlettae strain isolated from bovine mastitis milk led to the detection of a new 849-bp "bla$_{ARL}$ " gene, preceded by two regulatory genes, blaI$_{ARL}$ and blaR1$_{ARL}$, transcribed in reverse, forming a β-lactamase operon similar to the blaI- blaR-blaZ operon (Andreis et al., 2017).Target modification is mediated by mec gene products (2.1 kb). Resistance to meticillin, which leads to resistance to all β- lactams, is determined by the presence of a mecA chromosomal gene encoding a PLP2a transpeptidase. This additional PLP has less affinity for β- lactamins than other PLPs. This mechanism is present in methicillin-resistant Staphylococcus aureus (MRSA) and SCN. It has been proposed that this gene originated in coagulase-negative Staphylococcus species, such as S. sciuri or S. fleuretti, and that from these it was transferred to S. aureus (Quincampoix and Mainardi, 2001; Aguayo-Reyes et al., 2018). The mecA gene is located in a mobile genetic element: the staphylococcal cassette chromosome mec (SCCmec) inserted at a specific site on the chromosome: the attB site$_{scc}$, which lies next to the orfX gene encoding a ribosomal methyl transferase (Liu et al., 2016).Several types and variants of SCCmec are described on the basis of the different classes of mec complex and ccr gene allotypes. Generally, MRSA isolated from hospitals (MRSA-H) harbor SCCmec types I, II, III and sometimes IV, while those isolated from community infections (MRSA-CA) tend to harbor SCCmec types IV and V(Ouchenane et al., 2014).Chromosomal cassette variants all possess a mec gene complex, consisting of mecA (which codes for PBP2a) and the mecI and mecR genes, whose products regulate the expression of meticillin resistance. They also possess a ccr gene complex, composed of one or two genes that code for site-specific recombinases responsible for chromosomal cassette mobility (integration and excision phenomena) (Aguayo- Reyes et al., 2018). The SCCmec cassette also includes so-called "accessory" elements such as insertion sequences, transposons or plasmid copies carrying resistance genes to heavy metals and antibiotics other than β-lactamins (figure 9) (Dumitrescu et al., 2010).

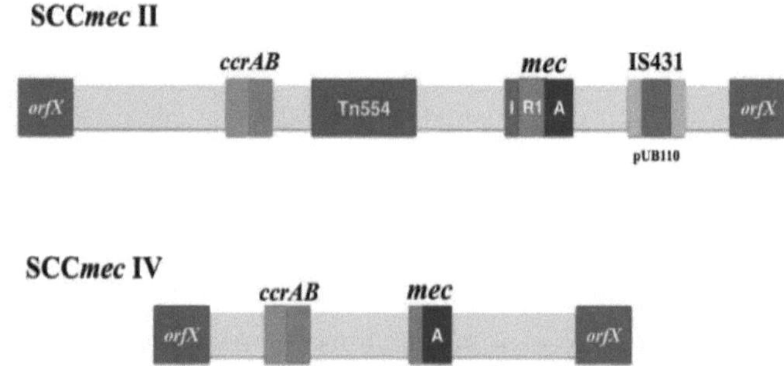

Figure 9. Simplified diagram of the general structure of the staphylococcal chromosomal cassette SCCmec II and IV (Aguayo-Reyes et al., 2018).

In 2011, a new variant of the mecA gene located on the new SCCmec XI cassette type was detected. It was named $_{mecALGA251}$ and then mecC. Seven years later, another meticillin resistance determinant (mecB) was described in Germany in an S. aureus isolated from a nasopharyngeal swab of a 67-year-old patient with no signs of infection (Becker et al., 2018; Ruiz-Ripa, 2020).

- β-lactamases

These enzymes irreversibly hydrolyze β-lactams by opening the β-lactam core via an unstable acyl-enzyme intermediate, ultimately leading to the loss of a carboxyl group (figure 10). They are excreted in the periplasmic space of Gram-negative bacteria (Poyart, 2003; Nauciel and Vildé, 2008; Ruppé, 2010).

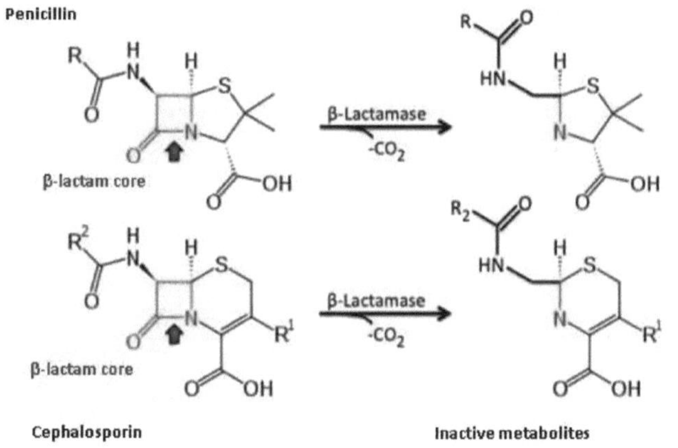

Figure 10: Enzymatic inactivation of β-lactam antibiotics (Nagshetty, 2021)

Two classification schemes for β-lactamases are currently in use:

Ambler molecular classification (Ambler, 1980): four groups (A, B, C and D) are distinguished according to their amino acid sequence and enzyme-substrate interaction mechanisms. Class A, C and D β-lactamases are known as serine β-lactamases (contain a serine in their active site), while class B enzymes are zinc-dependent metallo-β-lactamases (figure 11);

Bush, Jacoby and Medeiros functional classification: Based on the biochemical, structural and functional characteristics of β-lactamases, notably hydrolytic activity and sensitivity to inhibitors. It was proposed by Bush in 1985, revised by Bush, Jacoby and Medeiros in 1995, and updated by Bush and Jacoby in 2010. Enzymes are divided into three main groups (Bush et al., 1995; Bush and Jacoby, 2010):

Group 1 (Ambler class C): cephalosporinases not inhibited by β-lactamase inhibitors such as clavulanic acid and tazobactam (e.g. AmpC-type β-lactamases, CMY-2, FOX-1, etc.);

Group 2 (Ambler classes A and D): the largest group containing broad-spectrum enzymes, most of which are inhibited by clavulanic acid. These include most extended-spectrum β-lactamases (ESBLs) and serine carbapenemases (e.g. β-lactamases such as TEM-3, SHV-2, CTX-M, PER, VEB, KPC, OXA-23, OXA-48 etc.);

Group 3 (Ambler class B): metallo-enzymes inhibited by ethylenediaminetetraacetic acid (EDTA) (e.g. β-lactamases of type IMP-1, VIM-1, NDM-1),...etc.).

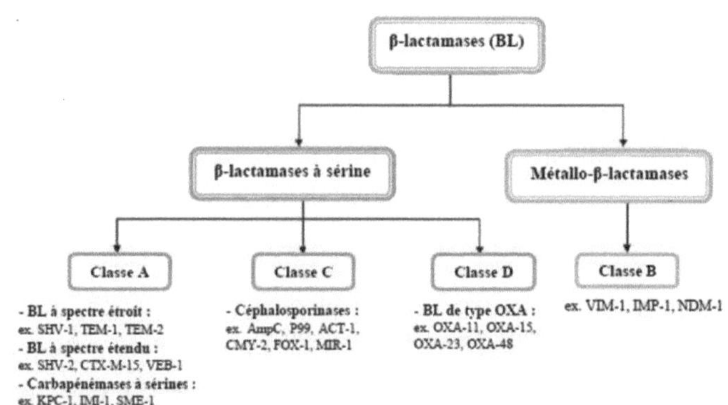

Figure 11. Classification scheme for β-lactamases according to Ambler (Vrancianu et al., 2020).

- Extended-spectrum β-lactamases (ESBL)

With the exception of OXA-type ESBLs (Ambler class D and Bush class 2), ESBLs are β-lactamases of Ambler class A and Bush-Jacoby- Medeiros class 2, i .e. they are capable o f hydrolyzing penicillins, anti-inflammatory drugs and other antibiotics. 1ère (C1G), 2ème (C2G), 3ème (C3G) (e.g. cefotaxime, ceftazidime) and 4ème (C4G) (e.g. cefepime) generation cephalosporins and monobactams (e.g. aztreonam). However, ESBL-producing strains generally remain sensitive to cephamycins (e.g. cefoxitin) and carbapenems (e.g. imipenem). Class A ESBLs are inhibited by β-lactamase inhibitors such as clavulanic acid, sulbactam and azobactam (Cattoir, 2008; Ruppé, 2010). The number of ESBL variants identified has grown steadily since 1983. At present, several dozen ESBL variants are known, and the list seems far from complete. It's worth mentioning that it was only in 1987 that the first CTX-M-type enzymes (specifically CTX-M-1) were reported in K. pneumoniae (Sirot et al., 1987).

- Former ESBLs (SHV and TEM types)

The first ESBL was detected in Germany in 1983 in a clinical strain of Klebsiella ozaenae. This β-lactamase was derived from the narrow-spectrum β-lactamase SHV-1, through a simple mutation. It was thus named SHV-2 (Kliebe et al., 1985). A few years later, other ESBL variants were identified (bla_{SHV2a}, bla_{SHV3}, bla_{SHV4} and bla_{SHV5}) showing variable genetic homologies ranging from 50 to 90% with the bla_{SHV-1} and bla_{SHV-2} sequences (Liakopoulos et al., 2016). The majority of SHV-type ESBLs are found in K. pneumoniae strains, but have also been detected in Citrobacter diversus, E. coli and P. aeruginosa (Bradford, 2001).Similarly, a potent cephalosporinase named CTX-1 was detected in several species of Enterobacteriaceae in Clermont-Ferrand, France. Subsequent sequencing showed that CTX-1 was a three-amino acid mutant of TEM-1, and the enzyme was thus renamed TEM-3 (Sirot et al., 1987). Although frequently found in E. coli and K. pneumoniae, TEM-type ESBLs have also been reported among other members of the Enterobacteriaceae family, as well as P. aeruginosa (Bradford, 2001). Today, SHV-type ESBLs account for just over 10% of ESBLs, mainly found in K. pneumoniae, while TEM-type ESBLs have become rare (Doi et al., 2017). As of July 2021, 243 TEM-type and 228 SHV-

type β-lactamases have been identified and listed in public databases   :
(https://www.ncbi.nlm.nih.gov/pathogens
/isolates#/refgene/TEM;https://www.ncbi.nlm.nih.gov/
pathogens/isolates#/refgene/SHV), although not all have the ESBL phenotype
(Castanheira et al., 2021).

- Main new ESBLs (CTX-M, OXA and PER types)

a. CTX-M type ESBL

CTX-M-type cephalosporinases are "emerging" enzymes that currently represent the most frequent ESBLs in enterobacteria worldwide, following their rapid spread since the late 1980s, a few years after the introduction of cefotaxime as a treatment for bacterial infections (Bonnet, 2004; Cattoir, 2008).The first detection of CTX-M was in Germany in early 1989, when Bauernfeind et al. (1996) identified a clinical strain of cefotaxime-resistant E. coli producing a non-TEM, non-SHV ESBL, designated CTX-M-1, in reference to its hydrolytic activity against cefotaxime. At the same time, an explosive spread of Salmonella strains resistant to this same antibiotic began in South America (Bauernfeind et al., 1992).Unlike TEM- and SHV-type ESBLs, which evolved through the accumulation of mutations in classical β-lactamases, CTX-M enzymes are derived from the chromosomal β- lactamase of various species of the genus Kluyvera spp. (Cantón et al., 2012). To date, over 172 types of CTX-M have been identified and described (https://www.lahey.org/studies/other.asp). They are grouped into six clusters: CTX-M-1, CTX- M-2, CTX-M-8, CTX-M-9, CTX-M- 25 and KLUC (Figure 12). Each cluster is named after the first member described and generally comprises minor allelic variants that differ from each other by a single or a few amino acids (Rossolini et al., 2008; Ramadan et al., 2019). The majority of variants are found in CTX-M groups 1 and 9, suggesting higher plasticity for these groups (D'Andrea et al., 2013). It has been proposed that the worldwide dissemination of bla$_{CTX-M}$ genes is due to several factors, including their location on mobile or mobilizable genetic elements, the easy co-selection resulting from their association with resistance genes against other clinically relevant antibiotics, and their presence in highly epidemic clones. The most relevant examples are represented by the pandemic E. coli clone ST131 (phylogenetic group B2), which greatly contributed to the worldwide dissemination of CTX-M-15, and by the E. coli clones of clonal complexes ST405 and ST38 (phylogenetic group D), which were associated with the emergence of the CTX-M-15 and CTX-M-9 and CTX-M-14 enzymes, respectively. In addition, E. coli ST10 (phylogenetic group A), which is a typical member of the human gut microbiota but also responsible for intestinal and extra-intestinal infections which is a typical member of the human gut microbiota but also responsible for intestinal infections has recently been associated with the spread of various groups including CTX-M- 1, CTX-M-2 and CTX-M-9 (Rogers et al., 2011; Naseer and Sundsfjord, 2011; D'Andrea et al., 2013).

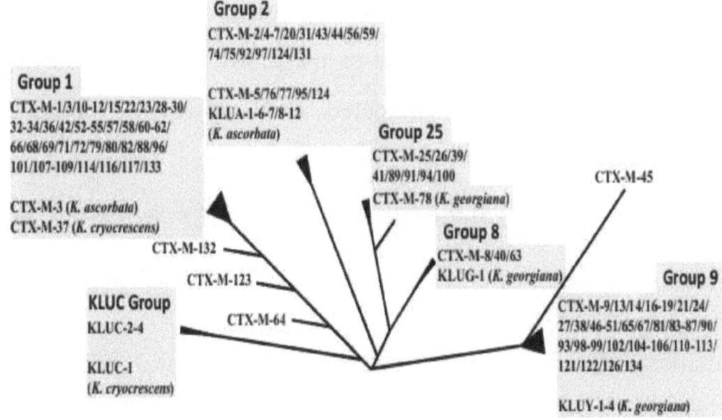

Figure 12. Tree diagram showing the different CTX-M enzyme clusters (D'Andrea et al., 2013)

b. OXA-type ESBL

The role of these class D enzymes remains secondary to the pandemic role of class A ESBLs. OXA-type β-lactamases confer resistance to ampicillin and cephalothin, and are characterized by their high hydrolytic activity against oxacillin and cloxacillin. However, they are weakly inhibited by clavulanic acid. OXA-type ESBLs were first detected in Turkey in 1991 in a clinical isolate of Pseudomonas aeruginosa that was multidrug-resistant, particularly to ceftazidime. After sequencing, the OXA-11 derivative, which differs from OXA-10 by two amino acids, was identified. Other ESBLs derived from OXA-10, namely OXA-13, OXA-14, OXA-16, OXA-17, OXA-19 and OXA-28, have also been described, particularly in P. aeruginosa (Brafdord, 2001; Evans and Amyes, 2014).

c. PER-type ESBL

The PER-1 enzyme, first discovered in 1993 in P. aeruginosa in Turkey, is common in P. aeruginosa and Acinetobacter spp. but has also been detected in S. enterica serovar Typhimurium, Providencia spp., Proteus mirabilis and Alcaligenes faecalis (Weldhagen et al., 2003). In Turkey, a recent study showed that 32% of ceftazidime-resistant strains of P. aeruginosa and 55% of those of A. baumannii were PER-1 producers (Naas et al., 2008). A second enzyme, PER-2 (86% identical to PER-1), was detected in 1996 in S. enterica serovar Typhimurium in Argentina, and since then in other enterobacteria, Vibrio cholerae and A. baumannii (Weldhagen et al., 2003; Naas et al., 2008).

3.2. Resistance to aminosides

Aminoglycosides" are powerful, broad-spectrum, natural or semi-synthetic antibiotics, mainly bactericidal through inhibition of protein synthesis. This class has been in clinical use since the discovery of streptomycin in 1944, isolated from Streptomyces griseus. Subsequently, other molecules were developed including neomycin (1949, S. fradiae), kanamycin (1957, S. kanamyceticus), gentamicin (1963, Micromonospora purpurea), tobramycin (1967, S. tenebrarius) and amikacin (1972, derived from kanamycin) (Krause et al., 2016).Three main mechanisms of resistance to aminoglycosides have been described in bacteria (figure 13) (Galimand et al., 2003; Doi et al., 2004;

Wachino et al., 2007):

1- Target modification due to ribosomal mutations or post-transcriptional
methylation of 16S RNA by ArmA, Rmt or NpmA methylases;

2- Antibiotic inactivation by enzymatic modification;

3- Decrease in intracellular antibiotic concentration.

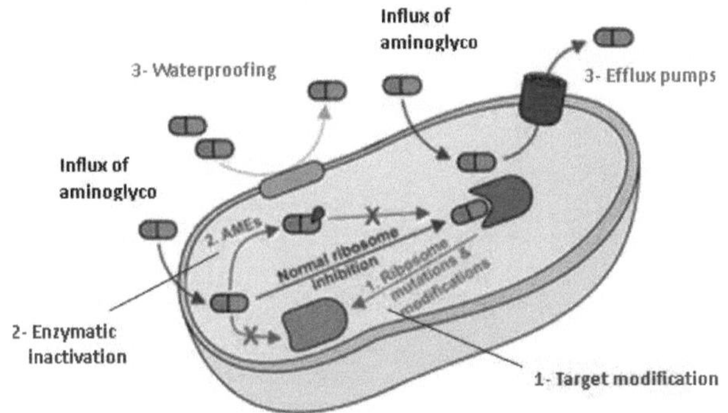

AMEs: aminoglycoside-modifying enzymes

Figure 13. Main mechanisms of resistance to aminoglycosides (Garneau-Tsodikova et al.Labby, 2015)

This latter mechanism remains the most frequently implicated in
aminoglycoside resistance in both Gram-negative and Gram-positive species.
Over 100 aminoglycoside-modifying enzymes (AMEs) have been described,
and are broadly classified into three groups according to their ability to
acetylate, phosphorylate or adenylate the amino or hydroxyl groups found in the
core structure of aminoglycosides (Mingeot-Leclerq et al., 1999):

- Acetyltransferases (AAC): by acetylation of an amine group ;

- Phosphotransferases (APH): by phosphorylation o f a hydroxyl group ;

- Nucleotidyltransferases (ANT): by nucleotidylation of a hydroxyl group (figure
14).

In S. aureus, resistance to these antibiotics is produced mainly by the action of enzymes encoded by the aac(6)-Ie-aph (2″)-Ia (bifunctional enzyme that confers resistance to gentamicin, tobramycin and kanamycin), ant (4)-Ia (confers resistance to tobramycin and kanamycin), aph (3′)-IIIa (confers resistance to kanamycin) and ant(6)-Ia genes. (confers streptomycin resistance) (Pascual, 2015).

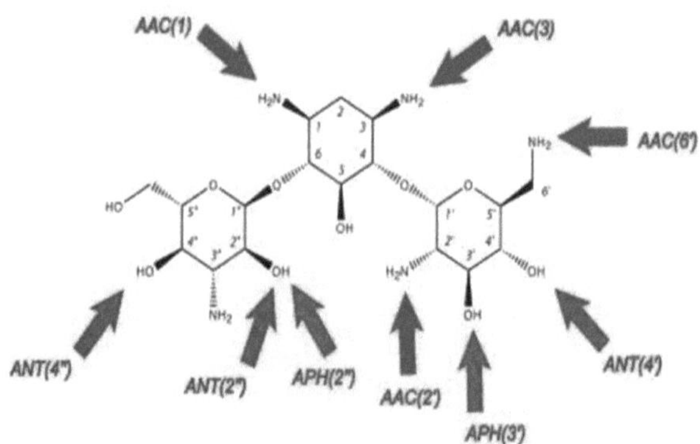

Figure 14. Examples of aminoglycoside modification sites by the enzymes AAC, APHand ANT for kanamycin (Krause et al., 2016).

In E. coli, this resistance is mainly mediated by the following genes aac(3)-I (confers resistance to gentamicin), aph(3″) (confers resistance to streptomycin), aph(3')-I and aph(3')-II (confers resistance to kanamycin and neomycin), aac(3)-II and aac(3)-IV (confers resistance to kanamycin, tobramycin and gentamicin) (Alonso, 2018).

3.3. Resistance to tetracyclines

Tetracyclines are well known for their broad spectrum of activity, covering a wide range of Gram-positive and Gram-negative bacteria, spirochetes, obligate intracellular bacteria and protozoan parasites. They are mainly bacteriostatic antibiotics, and exert their action by inhibiting protein synthesis by binding to the 30S subunit of ribosomes, preventing the entry of aminoacyl tRNA to the acceptor site of the mRNA-ribosome complex, and thus preventing the incorporation of new amino acids into the chain. polypeptide (Pioletti et al., 2001; Grossman, 2016). The first molecule, chlortetracycline, was isolated in 1948 from Streptomyces aurefaciens. Later, oxytetracycline was isolated from S. rimosus, and new molecules such as tetracycline, democlocycline, methacycline, doxycycline and minocycline were synthesized. The most recently discovered tetracyclines are the semi-synthetic group known as "glycylcyclines", of which "tigecycline" remains the most extensively studied and looks promising as a new antimicrobial agent that can be administered as monotherapy to patients with many types of serious bacterial infections (Chopra and Roberts, 2001; Noskin et al., 2005). Four main mechanisms of tetracycline resistance have been described in bacteria:

1- Antibiotic expulsion by efflux pumps;

2- Ribosomal protection;

3- Enzymatic inactivation of the antibiotic;

4- Mutations in 16S ribosomal RNA.

In enterobacteria, an important class of transporters is represented by the Tet A-E proteins responsible for tetracycline efflux from the cytosol to the periplasm (Mesaros et al., 2005). In S. aureus, the tet(K) and tet(L) genes promote active expulsion of the antibiotic. The second most important mechanism of tetracycline resistance is the expression of genes encoding ribosomal protection proteins, the most widespread of which are tet(O) and tet(M). The tet(K) gene is the most common in S. aureus. It confers resistance only to tetracycline, while the tet(M) gene can confer resistance to doxycycline and minocycline, in addition to tetracycline (Schmitz et al., 2001; Pascual, 2015).

3.4. Resistance to phenicols

Chloramphenicol, originally called chloromycetin, was isolated from Streptomyces venezuelae in 1947. It was the first natural substance described to contain a nitrophenyl group. Its relatively simple structure made it the first antibiotic to be marketed as a chemically synthesized product, and it has been produced exclusively in this way since 1950 (Schwartz et al., 2004).In the early years of its introduction into clinical therapy, chloramphenicol was considered a promising broad-spectrum antibiotic. Later, a number of adverse effects were observed: irreversible aplastic anemia, reversible bone marrow suppression, or Gray's syndrome in newborns and infants. Occasionally, hypersensitivity ranging from skin rashes to anaphylaxis has also been observed (Shaw, 1983; Yao et al., 1999). Chloramphenicol is a bacteriostatic agent that inhibits protein synthesis by binding to the 50S subunit of bacterial ribosomes, and blocks the peptidyl transferase reaction (Vinning and Stuttard, 1995). The primary mechanism of bacterial resistance to chloramphenicol, and still the most frequently encountered, is enzymatic inactivation by acetylation of the antibiotic by various types of chloramphenicol acetyltransferases (CATs). However, there are other resistance mechanisms, such as efflux systems (encoded mainly by the cmlA and floR genes in enterobacteria), inactivation by phosphotransferases, target mutations and decreased permeability (Shaw, 1983; Schwartz et al., 2004). The mechanisms most commonly detected in Staphylococcus spp. are mediated by the cat_{pc221} , cat_{pc223} , cat_{ps194} genes, and to a lesser extent by the cfr and fexA genes (Pascual, 2015). The cfr gene product (chloramphenicol and florfenicol resistance) is a methyltransferase that catalyzes the methylation of A2503 in the 23S rRNA gene of the large ribosomal subunit, conferring resistance to chloramphenicol, florfenicol, clindamycin and linezolid (Morales et al., 2010).

3.5. Resistance to quinolones

In 1962, the quinolones were discovered as an important treatment for various pathological manifestations. The first molecule in this class, nalidixic acid, was identified as a by-product of chloroquine synthesis and had limited clinical use, being sufficient only for the treatment of urinary tract infections. Over the years, this antibiotic has been modified to achieve a broader spectrum of activity (Gram-positive and Gram-negative) and better pharmacokinetic properties. The first fluoroquinolones (second-generation quinolones such as norfloxacin) appeared after the addition of a fluorine group to the central ring, generally in position 6.Currently, there are four generations of quinolones, including ofloxacin, ciprofloxacin, levofloxacin and moxifloxacin (Hooper and Jacoby, 2015).Quinolones act by forming a ternary complex between DNA and its two target enzymes: DNA gyrase (topoisomerases II) and topoisomerases. These enzymes have a similar tetrameric structure, composed of two A and two B subunits, respectively encoded by the gyrA and gyrB genes in the case of DNA gyrase, and byC andE in the case of topoisomerase IV (Jacoby, 2005). Topoisomerases are directly involved in the mechanisms of DNA uncoiling and supercoiling during DNA replication, transcription and recombination, in order to facilitate the action of DNA polymerase. Antibacterial activity in Gram-negative organisms mainly involves inhibition of DNA gyrase activity, while activity in Gram-positive organisms seems to involve blocking topoisomerase IV (Jacoby, 2005). - Three mechanisms of quinolone resistance are currently known (figure 15):

a- Chromosomal mutations that modify targets (topoisomerases) in the quinolone resistance determining region (QRDR): Among the various resistance-associated mutations described in this region, those occurring in codons Ser83 and Asp87 in GyrA, and Ser80 and Glu84 in ParC are the most frequently observed. However, the level of resistance varies according to the target altered and the number of amino acid substitutions accumulated.) In E.

coli, a single amino acid change in GyrA is sufficient to induce a high level of resistance to nalidixic acid and decreased susceptibility to fluoroquinolones. However, to obtain a high level of resistance to the latter, the presence of a second modification of GyrA and/or ParC is required (Ruiz, 2003);

b- Mutations that reduce quinolone accumulation through impermeability of the outer membrane (porin mutations) (Jacoby, 2005);

c- Mutations that induce overexpression of efflux pumps (e.g. AcrAB- TolC in E. coli and MexAB-OprM in Pseudomonas aurugenosa, etc.) (Jacoby, 2005);

d- Plasmid-mediated resistance genes: by topoisomerase protection of quinolone binding (qnr genes), efflux (QepA and OqxAB efflux pumps) and enzymatic inactivation by quinolone acetylation: (the AAC(6')-Ib-cr enzyme) (Minarini and Darini, 2012).

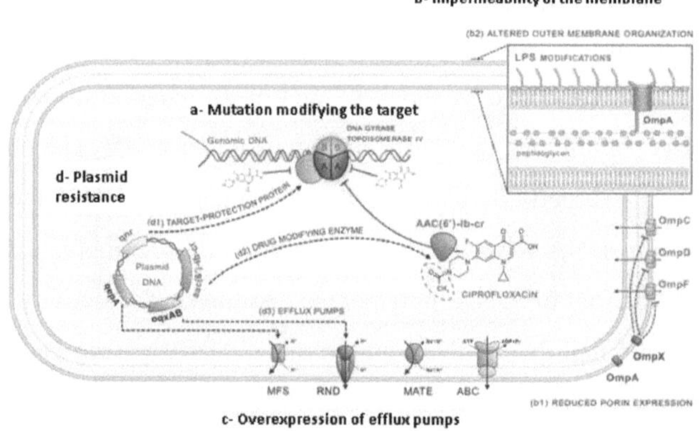

Figure 15. Mechanisms of quinolone resistance (Correia et al., 2017)

LPS: lipopolysaccharides; **MFS**: major animating superfamily; **RND**: resistance- nodulation - division; **MATE**: multiple extrusion of antibiotics and toxins; **ABC**: ATP-binding cassette; **MDR**: multidrug resistance.

3.6. Macrolide-Lincosamide-Streptogramin (MLS) resistance

Although different in chemical structure, these three groups o f antibiotics share the same mechanisms o f action and resistance. They inhibit bacterial protein synthesis at different levels of translocation, by binding to the bacterial peptide ribosome and thus inhibiting the subunit 50 S of the ribosome. In the group macrolide group and and lincosamide group, erythromycin and clindamycin are the most widely used antibiotics. Resistance toMLS ismainly due to three different mechanisms (figure 16)

1- Target modification due to the action of methylases, encoded by the erm and cfr genes;

2- Active antibiotic expulsion mediated by the genes msr(A)/msr(B), mef(A), mef(E),erp(B), vga(A), vga(B), vga(C), vga(D), vga(E), lsa(A), lsa(B), lsa(C), lsa(E) ;

3- Antibiotic inactivation by enzymes encoded by the lnu(A), lnu (B) and mph(C).

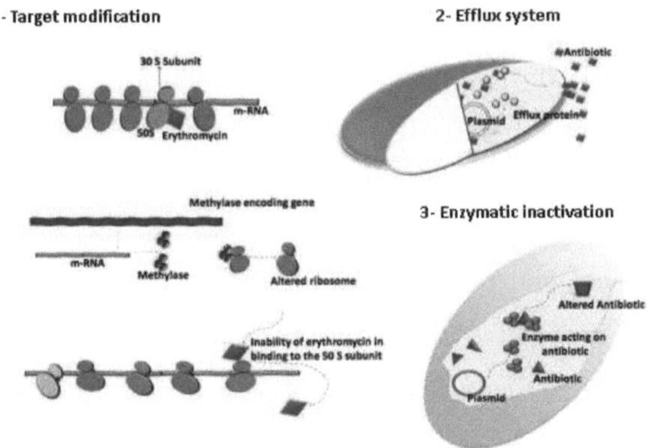

Figure 16. Mechanisms of bacterial resistance to MLS (Thumu et al. Halami, 2012)

3.7. Resistance to sulfonamides and trimethoprim

Sulfonamides were the first synthetic antimicrobial agents, first used clinically in 1935. They have been used for several decades as effective and inexpensive antibacterial agents in animals and humans. They are bacteriostatic and generally used in combination with diaminopyrimidines such as trimethoprim. The combination sulfamethoxazole Ŕ trimethoprim was marketed in 1968 under the international non-proprietary name "cotrimoxazole" (Sköld, 2000; Mosquito et al., 2011).Sulfonamides and trimethoprim affect the metabolism of folic acid, essential for nucleic acid synthesis. Sulfonamides competitively inhibit the enzyme dihydropteroate synthase (DHPS), due to their structural analogy with their natural substrate, p-aminobenzoic acid (PABA). Trimethoprim acts by inhibiting another enzyme, dihydrofolate reductase (RHFD), which oxidizes dihydrofolic acid to tetrahydrofolic acid (Mosquito et al., 2011). Bacterial resistance to sulfonamides occurs mainly due to mutations in the folP gene encoding dihydropteroate synthase (DHPS) involved in nucleotide biosynthesis, or through the acquisition of alternative mutated DHPS genes (sul1, sul2 and sul3), whose products have low affinity for sulfonamides (Kim et al., 2019). Recently, the fourth mobile sulfonamide resistance gene (sul4) has been shown to be widespread in Asia and Europe (Razavi et al., 2017). For trimethoprim, around twenty phylogenetically separated resistance genes expressing antibiotic-insensitive dihydrofolate reductases have been characterized (dfr genes). All these genes are efficiently propagated as cassettes in the variable region of integrons, transposons and plasmids (Sköld, 2000). In staphylococci, the most frequent resistance to trimethoprim is attributed to expression of the dfrG, dfrK, dfrA (also named dfrS1) and dfrD genes (Nurjadi et al., 2014; Reeve et al., 2016).

REFERENCES

A

Aguayo-Reyes, A. Quezada-Aguiluz M., Mella S., Riedel G., Opazo-Capurro A., Bello-Toledo H., Domínguez M., González-Rocha G. 2018. "Bases moleculares de la resistencia a meticilina en Staphylococcus aureus". Rev Chilena Infectol, 35(1):7-14.

Alekshun M.N., Levy S.B. 2007. "Molecular mechanisms of antibacterial multidrug resistance". Cell, 128(6):1037-50.

Alonso C.A. 2018. "Epidemiología molecular en Escherichia coli procedente de fauna salvaje: resistencia antimicrobiana, virulencia y diversidad genética". Doctoral thesis. Departamento de Agricultura y Alimentación, Universidad de La Rioja, Logroño. 378p.

Alton N., Vapnek D. 1979. "Nucleotide sequence analysis of the chloramphenicol. resistance transposon Tn9". Nature, 282: 864-869.

Ambler R.P. 1980. "The structure of beta-lactamases". Phil Trans R Soc Lond B Biol Sci, 289(1036): 321-331.

Andreis S.N., Perreten V., Schwendener S. 2017. "Novel β- Lactamase bla$_{ARL}$ in Staphylococcus arlettae." mSphere, 2(3):e00117-17.

B

Bauernfeind A., Casellas J. M., Goldberg M., Holley M., Jungwirth R., Mangold P., Rohnisch T., Schweighart S., Wilhelm R. 1992. "A new plasmidic cefotaximase from patients infected with Salmonella Typhimurium. Infection, 20:158-163.

Bauernfeind A., Stemplinger I., Jungwirth R., Ernst S., Casellas J. M.. 1996. "Sequences of β-lactamase genes encoding CTX-M-1 (MEN-1) and CTX-M-2 and relationship of their amino acid sequences with those of other β-lactamases". Antimicrob Agents Chemother, 40:509-513

Becker K., van Alen S., Idelevich E.A., Schleimer N., Seggewiß J., Mellmann A., Kaspar U., Peters G. 2018. "Plasmid-Encoded Transferable mecB- Mediated Methicillin Resistance in Staphylococcus aureus". Emerg Infect Dis, 24(2):242-248.

Bennett P.M. 2008. "Plasmid encoded antibiotic resistance: acquisition and transfer of antibiotic resistance genes in bacteria". Br J Pharmacol, 153(1):S347-357.

Brown H.J., Stokes H.W., Hall R.M. 1996. "The integrons In0, In2, and In5 are defective transposon derivatives. J Bacteriol; 178:4429-4437.

Bryskier A. 1999. "Antibiotics, antibacterial and antifungal agents. Editions Ellipses, 1216 p.

Bush K., Jacoby G.A., Medeiros A.A. 1995. "A functional classification scheme for beta-lactamases and its correlation with molecular structure". Antimicrob Agents Chemother, 39: 1211-1233.

Bush, K., Jacoby, G.A. 2010. "Updated functional classification of β-lactamases". Antimicrob Agents Chemother, 54: 969-976.

C

Cantón R., González-Alba J.M., Galán J.C. 2012. "CTX-M Enzymes: origin and diffusion". Front Microbiol, 3:110.

Carattoli A. 2008. "Animal reservoirs for extended spectrum β-lactamase producers". Clin Microbiol Infect, 14(1):117-23.

Carattoli A. 2013." Plasmids and the spread of resistance". Int J Med Microbiol, 303(6-7): 298-304.

Castanheira M, Simner PJ, Bradford PA. 2021. "Extended-spectrum β-lactamases: an update on their characteristics, epidemiology and detection". JAC Antimicrob Resist, 3(3):dlab092.

Cattoir V. 2008. "The new extended-spectrum β-lactamases (ESBLs)".

Pathologie Infectieuse en Réanimation, 204-209.

Chopra I., Roberts M. 2001. "Tetracycline antibiotics: mode of action, applications, molecular biology, and epidemiology of bacterial resistance". Microbiology and molecular biology reviews, 65(2): 232-260.

Clark C.A., Purins L., Kaewrakon P., Focareta T., Manning P.A. 2000. "The Vibrio cholerae O1 chromosomal integron". Microbiology (Reading), 146 (10):2605-2612.

Collis C.M., Kim M.J., Partridge S.R., Stokes H.W., Hall R.M. 2002. "Characterization of the class 3 integron and the site-specific recombination system it determines". Journal of bacteriology, 184(11), 3017-3026.

Correia S., Poeta P., Hébraud M., Capelo J.L., Igrejas G. 2017. "Mechanisms of quinolone action and resistance: where do we stand?" J Med Microbiol, 66(5):551-559.

Courvalin P. 2008. "Bacterial resistance to antibiotics: Combinations of biochemical and genetic mechanisms". Bull Acad Vét, 161(1): 6p.

Couturier M., Bex F., Bergquist P. L., Maas W.K. 1988. "Identification and classification of bacterial plasmids". Microbiological reviews, 52(3) : 375- 395.

Couvé-Deacon E. 2017. "Epidemiology and regulation of class 1 integrons in Acinetobacter baumannii". Thèse doctotale, University of Limoges, 212p.

D

D'Andrea M.M., Arena F., Pallecchi L., Rossolini G.M. 2013. "CTX-M-type β- lactamases: a successful story of antibiotic resistance". Int J Med Microbiol, 303: 305-317.

Desiderio P. 1954. "History of antibiotics". Revue d'histoire des sciences et de leurs applications, 7(2):124-138.

Doi Y., Iovleva A., Bonomo R.A. 2017. "The ecology of extended-spectrum β- lactamases (ESBLs) in the developed world". J Travel Med, 24:S44-S51.

Doi Y., Yokoyama K., Yamane K., Wachino J., Shibata N., Yagi T.,

Shibayama K., Kato H., Arakawa Y. 2004. "Plasmid-mediated 16S rRNA. methylase in Serratia marcescens conferring high-level resistance to aminoglycosides". Antimicrob Agents Chemother, 48: 491-496.

Dumitrescu O., Dauwalder O., Boisset S., Reverdy M.É., Tristan A., Vandenesch F. 2010. "Antibiotic resistance in Staphylococcus aureus: key points in 2010. Med Sci (Paris), 26(11):943-949.

E

Evans B.A. & Amyes S.G.B. 2014. "OXA β-lactamases." Clinical microbiology reviews, 27(2): 241-63.

F

Fong I.W., Shlaes D., Drlika K. 2019. "Antimicrobial Resistance in the 21st Century, Emerging Infectious Diseases of the 21st Century." 2nd edition, Springer: 774p.

Furuya E.Y., Lowy F.D. 2006. "Antimicrobial-resistant bacteria in the community setting". Nat Rev Microbiol, 4(1):36-45.

G

Galimand M., Courvalin P., Lambert T. 2003. "Plasmid-mediated high-level resistance to aminoglycosides in Enterobacteriaceae due to 16S rRNA methylation". Antimicrob Agents Chemother, 47: 2565-2571.

Garneau-Tsodikova S., Labby K.J. 2016. "Mechanisms of resistance to aminoglycoside antibiotics: overview and perspectives". Med Chem Comm, 7(1): 11-27.

Götz A., Pukall R., Smit E., Tietze E., Prager R., Tschäpe H., van Elsas J.D., Smalla K. 1996. "Detection and characterization of broad-host-range plasmids in environmental bacteria by PCR". Appl Environ Microbiol, 62(7):2621-8.

Grindley N.D., Joyce C.M. 1980. "Genetic and DNA sequence analysis of the kanamycin resistance transposon Tn903". Proc Natl Acad Sci, 77(12):7176- 80.

Grossman T.H. 2016. "Tetracycline Antibiotics and Resistance". Cold Spring Harb Perspect Med, 6(4):a025387.

Guérin E., Jové T., Tabesse A., Mazel D., Ploy M.C. 2011. "High-level gene cassette transcription prevents integrase expression in class 1 integrons". J Bacteriol, (20):5675-82.

Guillot J.F. 1989. "Appearance and evolution of bacterial resistance to antibiotics". Annales de Recherches Vétérinaires, INRA Editions, 20(1) : 3-16.

Gutmann L., Williamson R., Kitzis M.D., Acar J.F. 1986. "Synergism and antagonism in double beta-lactam antibiotic combinations". Am J Med, 80(5C):21-29.

H

Haniford D.B. 2006. "Transpososome dynamics and regulation in Tn10 transposition". Critical reviews in biochemistry and molecular biology, 41(6): 407-424.

Haren L., Ton-Hoang B., Chandler M. 1999. "Integrating DNA: transposases and retroviral integrases". Annu Rev Microbiol, 53, 245-281.

Helfand W.H. 1982. "History of antibiotics: The History of Antibiotics, a Symposium". History of Pharmacy, 70e year, 252(69-71).

Hochhut B., Lotfi .Y, Mazel D., Faruque S.M., Woodgate R., Waldor M.K. 2001. "Molecular analysis of antibiotic resistance gene clusters in Vibrio cholerae O139 and O1 SXT constins". Antimicrob Agents Chemother, 45(11):2991-3000.

Hooper D.C., Jacoby G.A. 2015. "Mechanisms of drug resistance: quinolone resistance". Annals of the New York Academy of Sciences, 1354(1): 12-31.

I

Jacoby G.A. 2005. "Mechanisms of Resistance to Quinolones." Clinical Infectious Diseases, 41(2): S120-S126,

Jouini A., Vinué L., Ben Slama K., Saenz Y., Klibi N., Hammami S., Boudabous A., Torres C. 2007. "Characterization of CTX-M and SHV extended-spectrum β-lactamases and associated resistance genes in Escherichia coli strains of food samples in Tunisia". J Antimicrob Chemother, 60:1137-1141.

K

Kim D.W., Thawng C.N., Lee K., Wellington E.M.H., Cha C.J. 2019. "A novel sulfonamide resistance mechanism by two-component flavin-dependent monooxygenase system in sulfonamide-degrading actinobacteria". Environ Int. 127:206-215

Kliebe C., Nies B.A., Meyer J.F., Tolxdorff-Neutzling R.M., Wiedemann B. 1985. "Evolution of plasmid-coded resistance to broad-spectrum cephalosporins". Antimicrob Agents Chemother, 28: 302-307.

Kobayashi S.D., Malachowa N., DeLeo F.R. 2014. "Pathogenesis of Staphylococcus aureus Abscesses". The American journal of pathology, 185(6): 1518-1527.

Kollek R., Oertel W., Goebel W. 1978. "Isolation and characterization of the minimal fragment required for autonomous replication ("basic replicon") of a copy mutant (pKN102) of the antibiotic resistance factor R1". Mol Gen Genet, 162(1):51-57.

Krause K.M., Serio A.W., Kane T.R., Connolly L.E. 2016. "Aminoglycosides: An Overview". Cold Spring Harbor perspectives in medicine, 6(6): a027029.

L

Leclercq R., Derlot R., Duval J. 1988. "Plasmid-mediated resistance to vancomycin and teicoplanin resistance in Enterococcus faecium". N Engl J Med, 319: 157-61.

Liakopoulos A., Mevius D., Ceccarelli D. 2016. "A Review of SHV Extended-Spectrum β-Lactamases: Neglected Yet Ubiquitous". Front Microbiol, 7:1374.

Liu J., Chen D., Peters B.M., Li L., Li B., Xu Z., Shirliff M.E. 2016. "Staphylococcal chromosomal cassettes mec (SCCmec): A mobile genetic element in methicillin-resistant Staphylococcus aureus". Microb Pathog, 101:56-67.

M

Mazel D. 2006. "Integrons: agents of bacterial evolution". Nat Rev Microbiol, 4(8):608-620.

Merlin C., Toussaint A. 1999. "Bacterial transposable elements". Med Sci, 15:8-9

Mesaros N., Van Bambeke F., Avrain L., Glupczynski Y., Vanhoof R. Plesiat P., Tulkens P.M. 2005. "Active efflux of antibiotics and bacterial resistance: state of the question and implications". La Lettre de l'Infectiologue : de la microbiologie à la clinique, 20 :117-126.

Minarini L.A., Darini, A.L. 2012. "Mutations in the quinolone resistance-determining regions of gyrA and parC in Enterobacteriaceae isolates from Brazil". Brazilian journal of microbiology, 43(4):1309-1314.

Mingeot-Leclercq M.P., Glupczynski Y., Tulkens P.M. 1999. "Aminoglycosides: activity and resistance". Antimicrobial agents and chemotherapy, 43(4): 727-737.

Morales G, Picazo J., Baos E., Candel F.J., Arribi A., Peláez B., Andrade R., De la Torre MA., Fereres J., Sánchez-García M. 2010. "Resistance to Linezolid is mediated by the cfr gene in the first report of an outbreak of

Linezolid-Resistant Staphylococcus aureus", Clinical Infectious Diseases, 50(6):821-825.

Mosquito S., Ruiz J., Bauer J.L., Ochoa T.J. 2011. "Mecanismos moleculares de resistencia antibiótica en Escherichia coli asociadas a diarrea". Rev Peru Med Exp Salud Pública, 28: 648-656.

Muylaert A. & Mainil J. 2013. "Bacterial resistance to antibiotics, mechanisms and their "contagiousness"". Annales de Médecine Vétérinaire, 156: 109-123.

N

Naas T., Poirel L., Nordmann P. 2008. "Minor extended-spectrum β-lactamases". Clin Microbiol Infect, 14 (1):42-52.

Nagshetty K., Shilpa B., Patil S., Shivannavar C., Manjula N. 2021. "An Overview of Extended Spectrum β-Lactamases and Metallo β-Lactamases". Advances in Microbiology, 11: 37-62.

Naseer U., Sundsfjord A. 2011. "The CTX-M conundrum: dissemination of plasmids and Escherichia coli clones". Microb Drug Resist, 1783-97.

Nauciel C., Vildé J.L. 2008. "Medical bacteriology. 2^{éme} editions". Editions Masson: 272p.

Nesvera J., Hochmannová J., Pátek M. 1998. "An integron of class 1 is present on the plasmid pCG4 from Gram-positive bacterium Corynebacterium glutamicum". FEMS Microbiol Lett, 169(2):391-5.

Nield B.S., Holmes A.J., Gillings M.R., Recchia G.D., Mabbutt B.C., Nevalainen K.M., Stokes H.W. 2001. "Recovery of new integron classes from environmental DNA". FEMS Microbiol Lett, 195(1):59-65.

Noskin G.A. 2005. "Tigecycline: A new Glycylcycline for treatment of serious snfections". Clinical Infectious Diseases, 41(5):S303-S314.

Nurjadi D., Olalekan A. O., Layer F., Shittu A. O., Alabi A., Ghebremedhin B., Zanger P. 2014. "Emergence of trimethoprim resistance gene dfrG in Staphylococcus aureus causing human infection and colonization in sub-Saharan

Africa and its import to Europe". Journal of Antimicrobial Chemotherapy, 69(9):2361-2368.

O

Ouchenane Z., Agabou A., Smati F., Rolain J., Raoult M. D. 2013. "Staphylococcal cassette chromosome mec characterization of methicillin-resistant Staphylococcus aureus strains isolated at the military hospital of Constantine/Algeria", 61(6): 280-281.

P

Pagès J . M. 2004. "Bacterial porins and sensitivity to antibiotics". médecine/sciences, 20(3): 346-351.

Pascual D.B. 2015. "Líneas genéticas, virulencia y resistencia a antibióticos en Staphylococcus aureus de diferentes orígenes". Doctoral thesis, Logroño. Universidad de La Rioja, 320p.

Paterson DL., Bonomo RA. 2005. "Extended-spectrum β-lactamases: a clinical update". Clin Microbiol Rev,18(4):657-86.

Pioletti M., Schlünzen F., Harms J., Zarivach R., Glühmann M., Avila H., Bashan A., Bartels H., Auerbach T, Jacobi C., Hartsch T., Yonath A., Franceschi F. 2001. "Crystal structures of complexes of the small ribosomal subunit with tetracycline, edeine and IF3". EMBO J, 20(8):1829- 39.

Poyart C., Jardy L., Quesne G., Berche P., Trieu-Cuot P. 2003. "Genetic basis of antibiotic resistance in Streptococcus agalactiae strains isolated in a French hospital". Antimicrobial agents and chemotherapy, 47(2): 794-797.

Q

Quincampoix J.C., Mainardi J.L. 2001. "Mechanisms of resistance in Gram-positive cocci". Réanimation, 10(3): 267-275.

R

Rådström P., Sköld O., Swedberg G., Flensburg J., Roy P.H., Sundström L. 1994. "Transposon Tn5090 of plasmid R751, which carries an integron, is related to Tn7, Mu, and the retroelements". J Bacteriol, 176(11):3257-68.

Ramadan A.A., Abdelaziz N.A., Amin M.A., Aziz R.K. 2019. "Novel bla$_{CTX-M}$ variants and genotype-phenotype correlations among clinical isolates of extended spectrum β-lactamase-producing Escherichia coli". Sci Rep, 9(1):4224.

Razavi M. , Marathe N.P. Gillings M.R. , Flach C.F. , Kristiansson E. , Larsson D.G. J. 2017. "Discovery of the fourth mobile sulfonamide resistance gene" Microbiome, 5: 160.

Reeve S. M., Scocchera E. W., G-Dayanadan N., Keshipeddy S., Krucinska J., Hajian B., Ferreira J., Nailor M., Aeschlimann J., Wright D. L., Anderson A. C. 2016. "MRSA isolates from United States hospitals carry dfrG and dfrK resistance genes and succumb to propargyl-linked antifolates". Cell chemical biology, 23(12): 1458-1467.

Rogers B.A., Sidjabat H.E., Paterson D.L. 2011. "Escherichia coli O25b-ST131: a pandemic, multiresistant, community-associated strain". J Antimicrob Chemother, 66:1-14.

Rossolini G.M., D'Andrea M.M., Mugnaioli C. 2008. "The spread of CTX-M-type extended-spectrum -lactamases". Clin Microbiol Infect 14 (1): 33-41.

Ruiz J. 2003. "Mechanisms of resistance to quinolones: target alterations, decreased accumulation and DNA gyrase protection". J Antimicrob Chemother, 51: 1109-1117.

Ruiz-Ripa L. 2020. "Epidemiología molecular de Staphylococcus spp. desde un enfoque One Health: genes emergentes e inusuales de resistencia a antibióticos y de virulencia". Doctoral thesis, Logroño. Universidad de La Rioja, 298p.

Ruppé E. 2010. "Epidemiology of extended-spectrum β-lactamases: the advent. CTX-M". Antibiotics, 12(1): 3-16.

S

Sabaté M. & Prats G. 2002. "Estructura y función de los integrones". Enferm Infecc Microbiol Clin, 20(7):341-5.

Schmitz F.J., Krey A., Sadurski R., Verhoef J., Milatovic D., Fluit A.C.; European SENTRY Participants. 2001. "Resistance to tetracycline and distribution of tetracycline resistance genes in European Staphylococcus aureus isolates". J Antimicrob Chemother, 47(2):239-40.

Shaw W.V. 1983. "Chloramphenicol acetyltransferase, enzymology and molecular biology". Crit Rev Biochem, 14: 1-46.

Shintani M., Sanchez Z.K., Kimbara K. 2015. "Genomics of microbial plasmids: classification and identification based on replication and transfer systems and host taxonomy." Front Microbiol, 6:242.

Siguier P., Gourbeyre E., Varani A., Ton-Hoang B., Chandler M. 2015. "Everyman's Guide to Bacterial Insertion Sequences". Microbiol Spectr, 3(2):MDNA3-0030.

Sirot D., Sirto J., Labia R., Morand A., Courvalin P., Darfeuille-Michaud A., Perroux R., Cluzel R. 1987. "Transferable resistance to third-generation cephalosporins in clinical isolates of Klebsiella pneumoniae: identification of CTX-1, a novel β-lactamase". J Antimicrob Chemother, 20: 323-334.

Sköld O. 2000. "Sulfonamide resistance: mechanisms and trends" Drug Resistance Updates, 3(3):155-160.

Skurnik D. 2009. "Integrons: Structure and epidemiologyIntegrons: Structure and epidemiology". 11(2):116-129.

Suárez C., Gudiol F. 2009. "Antibióticos betalactámicos". Enferm Infecc Microbiol Clin, 27(2):116-129.

Sun J., Deng Z., Yan A. 2014. "Bacterial multidrug efflux pumps: mechanisms, physiology and pharmacological exploitations". Biochem Biophys Res Commun, 453(2):254-67.

T

Tang S.S., Apisarnthanarak A., Hsu L.Y. 2014. "Mechanisms of β-lactam antimicrobial resistance and epidemiology of major community- and healthcare-associated multidrug-resistant bacteria". Adv Drug Deliv Rev, 30(78):3-13.

Thumu S.C., Halami P.M. 2012. "Acquired resistance to macrolide-lincosamide- streptogramin antibiotics in lactic Acid bacteria of food origin". Indian journal of microbiology, 52(4):530-537.

U

Uttley A.H., George R.C., Naidoo J., 1989. "High-level vancomycin-resistant enterococci causing hospital infection". Epidemiol Infect, 103: 173-81.

V

Varani A., He S., Siguier P. 2021. "The IS6 family, a clinically important group of insertion sequences including IS26". Mobile DNA, 12: 11.

Vinning L.C., Stuttard C. 1995. "Chloramphenicol. Biotechnology, 28: 505 - 530.

Vinué L., Sáenz Y., Rojo-Bezares B., Olarte I., Undabeitia E., Somalo S., Zarazaga M., Torres C. 2010. "Genetic environment of sul genes and characterisation of integrons in Escherichia coli isolates of blood origin in a Spanish hospital". Int J Antimicrob Agents, 35: 492-496.

Vrancianu C.O., Gheorghe I., Dobre E.G., Barbu I.C., Cristian R.E., Popa M., Lee S.H., Limban C., Vlad I.M., Chifiriuc M.C. 2020. "Emerging Strategies to Combat β-Lactamase Producing ESKAPE Pathogens". Int J Mol Sci, 21(22):8527.

W

Wachino J., Shibayama K., Kurokawa H., Kimura K., Yamane K., Suzuki S., Shibata N., Ike Y., Arakawa Y. 2007. "Novel plasmid-mediated 16S rRNA m1A1408 methyltransferase, NpmA, found in a clinically isolated Escherichia coli strain resistant to structurally diverse aminoglycosides". Antimicrob Agents Chemother 51: 4401-4409.

Wang S., Dai E., Jiang X., Zeng L., Cheng Q., Jing Y., Hu L., Yin Z., Gao B., Wang J., Duan G., Cai X., Zhou D. 2019. "Characterization of the plasmid of incompatibility groupsIncFIIpKF727591 and IncpKPHS1 from Enterobacteriaceae species". Infectionand drug resistance, 12: 2789-2797.

Weldhagen G.F., Poirel L., Nordmann P. 2003. "Ambler class A extended-spectrum β-lactamases in Pseudomonas aeruginosa: novel developments and clinical impact". Antimicrob Agents Chemother, 47(8):2385-92.

Wright G.D. 2010. "Antibiotic resistance in the environment: a link to the clinic?" Curr Opin Microbiol, (5):589-94.

Y

Yao, J.D.C., Moellering Jr, R.C.. Murray, P.R., Baron, E.J., Pfaller, M.A., Tenover, F.C., Yolken, R.H., 1999. "Manual of Clinical Microbiology. 7th edn. ASM Press, Washington, D.C. 1474-1505.

Printed by Books on Demand GmbH, Norderstedt / Germany